W9-BYX-923

DISCOVERING MATH

ADDITION

DAVID L. STIENECKER

ART BY RICHARD MACCABE

BENCHMARK BOOKS

MARSHALL CAVENDISH

NEW YORK

Benchmark Books
Marshall Cavendish Corporation
99 White Plains Road
Tarrytown, New York 10591-9001

©Marshall Cavendish Corporation, 1996

Series created by Blackbirch Graphics, Inc.

Printed and bound in the United States.

Library of Congress Cataloging-in-Publication Data

Stienecker, David.
    Addition / by David Stienecker; illustrated by Richard Maccabe.
        p.   cm. — (Discovering math)
    Includes index.
    ISBN 0-7614-0593-3 (lib. bdg.)
    1. Addition—Juvenile literature. 2. Mathematical recreations—Juvenile literature. [1. Addition. 2. Mathematical recreations.] I. Maccabe, Richard, ill. II. Title. III. Series.
QA115.S783   1995
513.2'11—dc20
                                      95-13574
                                        CIP
                                        AC

Contents

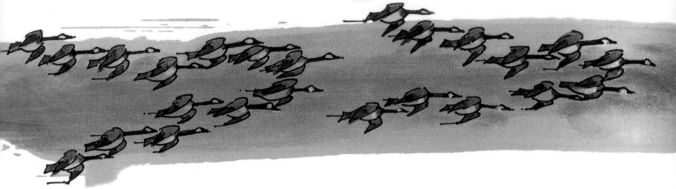

Birds on the Wing

Each fall, flocks of geese fly south for the winter. The geese form a V as they fly. A lead goose forms the point of the V. You can hear the geese honking as they fly.

> Combining two or more groups of things is called addition.

• Imagine that you look up at the sky and see these two flocks of geese. How many geese are in each flock? Write each number on a sheet of paper so you can remember it.

• The two flocks decide to form one flock. Draw a picture to show what the large flock looks like.

• How many geese are in your picture? How did you figure out how many geese to draw?

• Four lost geese come along and join the flock. Add them to your picture. How many geese are in the flock now?

• The geese get tired and hungry from all that flying. So they land on a pond to rest and look for food. There were already twenty geese on the pond. How many geese are there now?

Be a Hat Checker

Did you ever sit around wondering if more people wear hats than don't? Probably not. But maybe it's time you did. What's your prediction?

Do a hat survey to find out. Include all kinds of hats—all shapes, sizes, and colors.

Make a chart like this one on a sheet of paper. Use tally marks to record what people are wearing.

Hats	No Hats
~~IIII~~ I	IIII

Next you'll need a place to watch people's heads. Get a friend to help if you want.

Stick with it. The more heads you see, the better your survey will be. Do it until you've had it with hats. Then count up your tally marks.

• How many people were wearing hats? How many people weren't wearing any hats?

• Was your prediction right?

• How many heads did you count in all?

• What's that? You say you've had it with hats? Try eyeglasses, sneakers, shoelaces, or blue jeans. The possibilities are endless.

Tally charts make it easy to keep track of things. Each / is one. Each ~~IIII~~ is five.

5

Number Shapes

You can arrange numbers in shapes so their sums are the same. Look at this number shape. It looks like a triangle. Add up each side of the triangle. What is each side's sum?

You can add two or more numbers in any order. The sum is always the same.

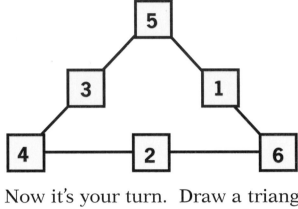

• Now it's your turn. Draw a triangle like the one above. Find three different ways to arrange the numbers 1, 2, 3, 4, 5, and 6 so each side equals 12. Remember to use each number only once.

• Now draw some more triangles. See if you can arrange the numbers 1, 2, 3, 4, 5, and 6 so each side equals 9. There is more than one solution. Work with a friend. See who can find a solution first.

• Try the same triangle with the even numbers 2, 4, 6, 8, 10, and 12. Make the sides add up to 18 and then 24.

• Try the same triangle with the odd numbers 1, 3, 5, 7, 9, 11. Make the sides add up to 15 and then 21.

Here's a larger triangle. How many addends make each side? (An addend is any number that can be added to another.) What is the sum of each side of the triangle?

• Copy this triangle on a piece of paper. Then see if you can use the numbers 1, 2, 3, 4, 5, 6, 7, 8, and 9 to make each side of the triangle equal 20 and then 17.

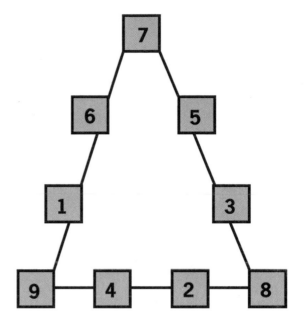

Here's another number shape. It looks like a wheel. All of the numbers going down and diagonally have the same sum. What is it?

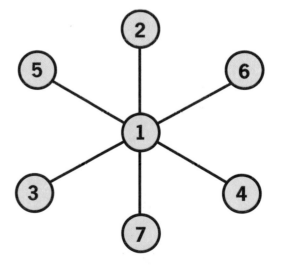

• Draw two number wheels on a sheet of paper. Then arrange the numbers 1, 2, 3, 4, 5, 6, and 7 so that each row adds up to 12 and then 14. Hint: Use a 4 in the center of one wheel and a 7 in the center of the other.

7

Hit the Target

It seems like everyone has a calculator. But calculators haven't been around all that long. It was only about forty years ago that calculators became popular.

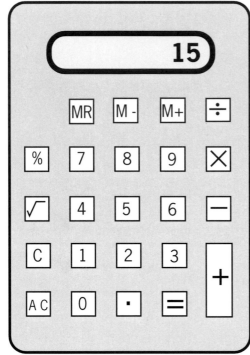

Try playing this calculator game with a friend.

1. First press the C button to clear the calculator's memory.

2. Decide on a target number. For your first game you might want to think small.

3. Take turns entering a number from 1 through 5. Enter each number by pressing the M+ button. Each new sum is put into the calculator's memory so you can't see it.

4. A player who thinks he or she has reached the target number presses the MR button to check. But beware. Once the sum appears, your partner has an advantage.

Here are some other ways to play the game.

• Choose larger and larger target numbers.

• Enter numbers 1 through 9.

• Play with three or four players. The more players, the harder it gets.

The Shortest Way 'Round

Draw or trace this rectangle onto a sheet of paper. Then cut it out and find its perimeter.

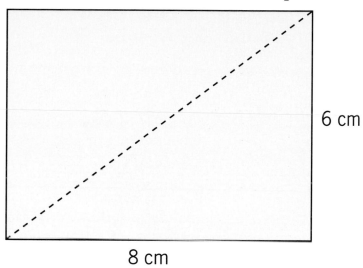

6 cm

8 cm

The perimeter is the distance around a figure. Just add up the lengths of all the sides.

Cut the rectangle in half along the dotted line. What two geometric shapes do you have now? What is the perimeter of each one?

• You can use your two triangles to make many different geometric shapes. Try making these. Use a metric ruler to figure out the perimeter of each one. Keep track of the shapes and their perimeters on a sheet of paper.

 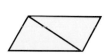 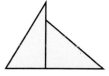

• Now for the big question. Which of your geometric shapes had the shortest perimeter? Which had the longest?

9

Pyramid Magic

These are special number pyramids. The numbers are arranged so you can do something that will really astound your friends.

You can use these pyramids. Or really impress everyone by making a large set of your own.

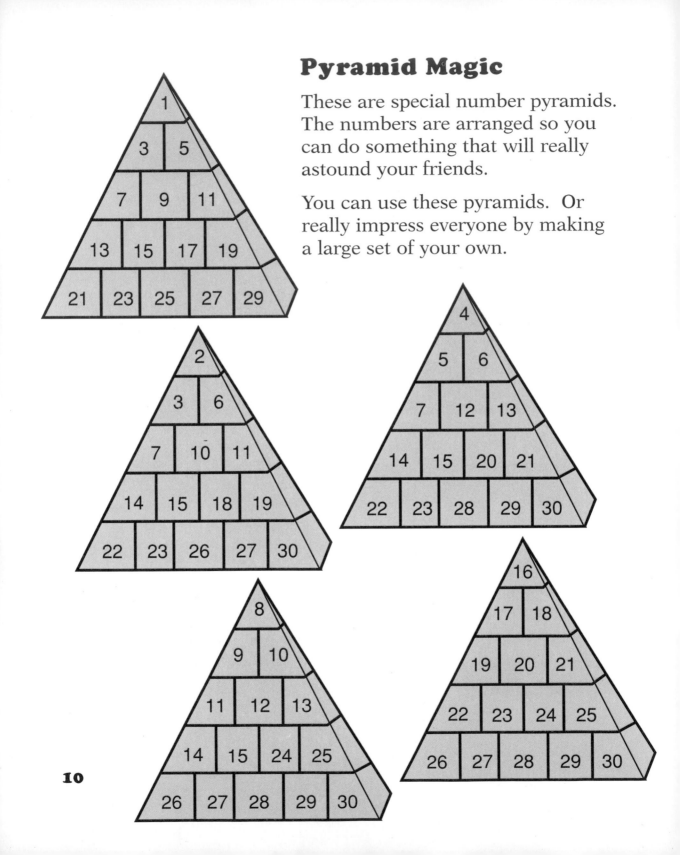

Here's what to do:

• Ask someone to think of a number 1 through 30.

• Have them place a coin, paper clip, or some other marker on each pyramid where their number appears.

• Stare deeply at the pyramids for several seconds. Mumble something about the pyramids sending you messages.

• With a big flourish, announce the person's number.

The magic in this trick is very simple. All you do is add up the numbers at the top of each pyramid the person identified. It works every time.

Here's an example just to convince you. Suppose you choose the number 13. These are the pyramids in which the number 13 appears.

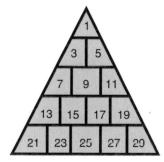 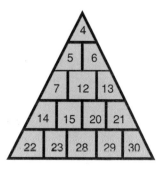 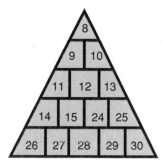

Now look at what happens when you add the numbers at the top of each pyramid:
1 + 4 + 8 = 13!

Magic in a Square

For thousands of years people thought numbers had magic powers. When certain numbers were arranged into magic squares they became even more powerful. Many people wore magic squares for good luck and to protect themselves from evil.

The oldest known magic square was found in an ancient Chinese book called the *I Ching*. The book was written over 4,000 years ago. This is what it looked like. Can you tell what its secret is?

Give up? Here are some clues. Add the numbers in each row and column. Add the numbers in each diagonal. What are the sums?

Squares like this are called magic squares. There are two rules that make them magic:

1. The sum of the numbers in each row, column, and diagonal must be the same.

2. Each number is used only once.

Here are two magic squares with numbers missing. Copy them on a piece of paper. Add the missing numbers.

8	1	
3		
4	9	

		10
	9	
8		6

Make up two more magic squares by adding 6 to each number in the magic squares above. Add different numbers to make more magic squares.

A SUPER MAGIC SQUARE

This magic square is made up of 16 numbers. Can you fill in the missing ones? Try using this magic square to make other magic squares in the same size.

7		1	14
	13		
16		10	
9	6		4

• Here's one more secret about magic squares. The numbers form a pattern. Can you figure it out? Give up? Here's a clue. Write down all the numbers in any magic square from the lowest number to the highest number.

1	2	3	4	5	6	7
8	9	10	11	12	13	14
15	16	17	18	19	20	21
22	23	24	25	26	27	28
29	30	31	32	33	34	35
36	37	38	39	40	41	42

Four by Four

Here's a way to amaze your friends. First you need to make a number grid like this one, only bigger. (Actually, this is just an example. Any number grid will do.)

Show the number grid to someone. Ask them to draw a box around any four-by-four block of numbers like this.

1	2	3	4	5	6	7
8	9	10	11	12	13	14
15	16	17	18	19	20	21
22	23	24	25	26	27	28
29	30	31	32	33	34	35
36	37	38	39	40	41	42

You can double a number by adding it to itself. Or, you can multiply it by two.

As the person is drawing the box, look at the pairs of numbers in opposite corners (3 and 27, or 6 and 24). Add either pair together and double the sum (60). Write your answer on a piece of paper and put it in your pocket. Then give these instructions.

Circle one of the boxed-in numbers. Then cross out all the other numbers in both the circled number's row and column.

1	2	3	4	5	6	7
8	9	10	11	14
15	16	17	18	19	20	21
22	23	24	25	26	27	28
29	30	31	32	33	34	35
36	37	38	39	40	41	42

14

Circle another number that isn't circled or crossed off. Then cross out all the other numbers in that number's row and column.

1	2	3	4	5	6	7
8	9	10	(11)	12	13	14
15	16	17	18	19	20	21
22	23	24	25	(26)	27	28
29	30	31	32	33	34	35
36	37	38	39	40	41	42

Repeat for a third number.

1	2	3	4	5	6	7
8	9	10	(11)	12	13	14
15	16	(17)	18	19	20	21
22	23	24	25	(26)	27	28
29	30	31	32	33	34	35
36	37	38	39	40	41	42

Circle the one number that is left.

1	2	3	4	5	(6)	7
8	9	10	(11)	12	13	14
15	16	(17)	18	19	20	21
22	23	24	25	(26)	27	28
29	30	31	32	33	34	35
36	37	38	39	40	41	42

Have the person add up the circled numbers. Then whip out your slip of paper and read the number on it. It will match the sum of the circled numbers!

You can do this trick with any number grid that is larger than four by four. You can start with any number if the numbers are consecutive.

• Try the trick with different-sized number grids. But always have the person box off a four-by-four block.

Coin Teasers

The first coins were made about 2,700 years ago in eastern Turkey. The first coins in the United States were made in 1652 by the Massachusetts Bay Colony. Here are some coin teasers to test your money savvy.

These coins are worth 30¢. See how many other combinations of coins you can come up with that are also worth 30¢. You can use any combination of quarters, dimes, nickels, and pennies.

You can work out your combinations on a sheet of paper. Better yet, use real coins. You will need 1 quarter, 3 dimes, 6 nickels, and 30 pennies to make all the combinations. A chart like this will help you keep track.

quarters	0										
dimes	2										
nickels	2										
pennies	0										

• Play a game with a friend. Take turns finding new combinations. The person who finds the most combinations wins.

• Try other amounts such as 25¢ or 35¢. But be warned. The higher the amount the more combinations there are likely to be.

Try this coin teaser. You must pay each amount
on the price tags. Figure out two combinations
of coins you can use. Draw the chart on a
sheet of paper to help you keep track of your
combinations. The first one has been done.

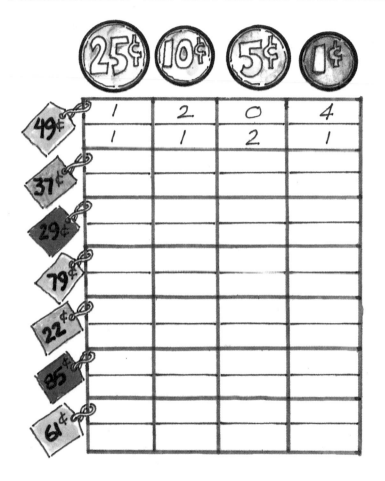

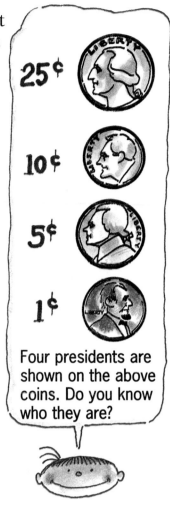

Four presidents are
shown on the above
coins. Do you know
who they are?

• Try finding three combinations. Then four.
Continue finding combinations until you're tired
of the game.

• The next time you buy something, see how many
different combinations of coins you can use.

Sum Game!

Here's an addition game you can play with any number of people. These are things you need:

a deck of playing cards

a pencil for each player

a piece of paper for each player

Before playing, prepare the deck by taking out all the face cards and ten cards. Let the aces stand for 1's and the jokers stand for 0's.

hundreds	tens	ones

Then each player should draw a game board like this one on a sheet of paper.

How to play:

1. Shuffle the cards and lay them face down.

2. In turn, each player takes a card and reads the number aloud. Remember aces are 1's and jokers are 0's.

3. Each player writes the number in any of the boxes in the top two rows of his or her game board. Here's how one player's game board might look after a seven and two card have been drawn.

hundreds	tens	ones
		2
	7	

4. After six cards have been drawn, each player's game board will have two three-digit numbers. This is how a game board might look at this stage of the game.

hundreds	tens	ones
5	1	2
4	7	9

5. Each player should add the numbers on his or her game board. The winner is the person with the largest sum.

• What is the sum of this player's game board? How could the digits be arranged differently to make a larger sum? Give it a try. See how the game works!

• Try this twist on the game. After everyone has added their numbers, give them sixty seconds to rearrange the digits to make a larger sum.

• Here's another way to play the game. Have players add a thousands column to their game boards. Draw eight cards instead of six.

Zoo Maze

You might be handed a map like this one when you visit a zoo. It shows all the paths to the different animals. This map also shows how long each path is in meters (m).

Use the map to plan these mini tours. Figure out the shortest distance from one place to the other:

• from the entrance to the monkeys

• from the monkeys to the elk

• from the elk to the gorillas

• from the gorillas to the seals

• from the seals to the camels

• from the camels to the entrance

Now plan a tour of the entire zoo! It's almost like finding your way through a maze. Start and end at the entrance. Visit each animal exhibit once. Never use the same path twice. Find out how long your tour is. Use a calculator to keep track of the meters you travel.

• Try finding another route for your tour. Keep track of the meters traveled. Is this tour shorter or longer than the one you made before?

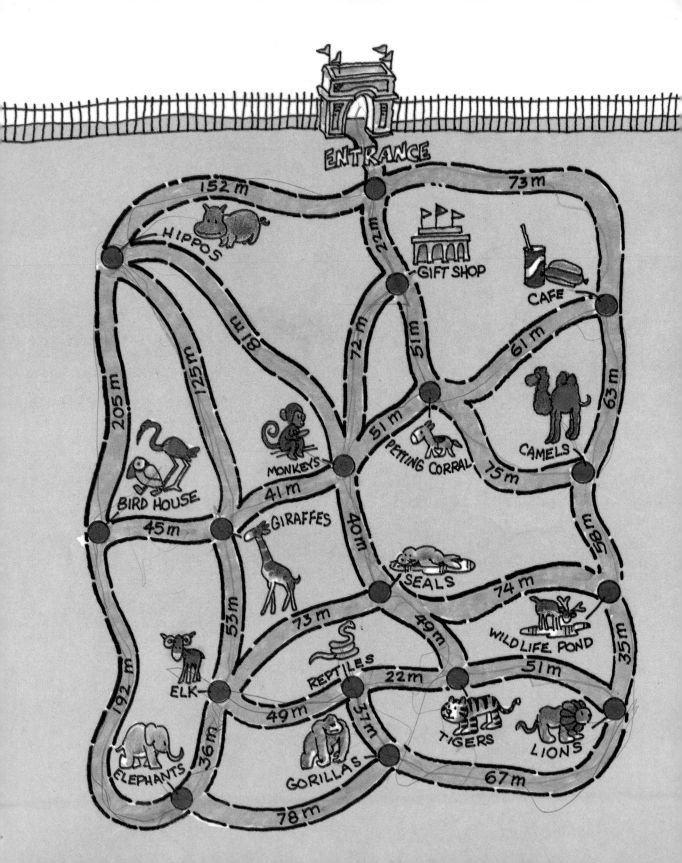

Dinosaur Trek

Play this game with a friend and trek through the land of dinosaurs. Use a penny and a dime for markers. Throw one die (one of a pair of dice) to tell how many spaces to move. (If you don't have dice, make six number cards. Draw and replace a card during each turn. Mix the cards up between turns.)

How to play:

1. The first player throws the die, then moves that number of spaces. Count the "Start" square.

2. If you land on a blue sum square, you don't have to do anything until your next turn. If you land on a square with an addition problem, find the sum and move your marker to the square where the sum appears. If you land on a square with words, do what it says.

3. Watch for those short cuts and danger zones! They can make life very easy or really difficult.

4. The first person to get to the end wins!

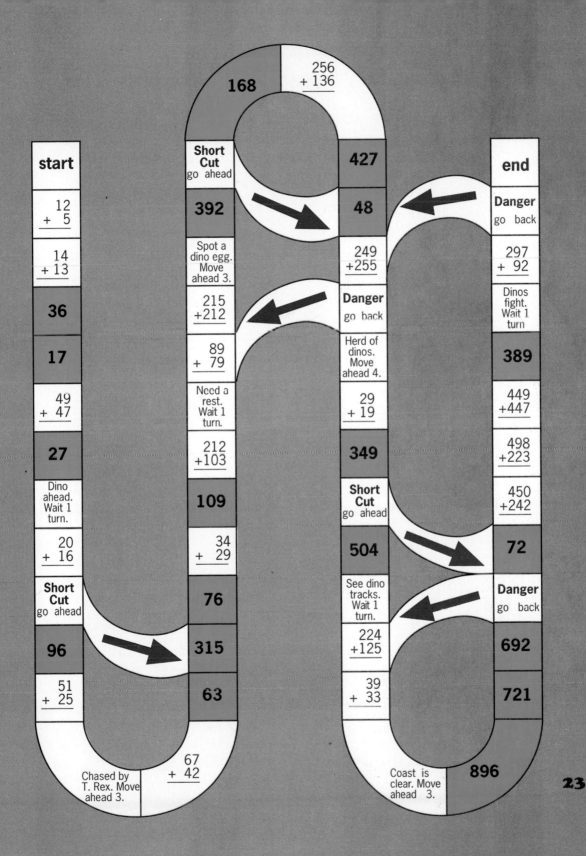

start

$\begin{array}{r} 12 \\ + 5 \\ \hline \end{array}$

$\begin{array}{r} 14 \\ + 13 \\ \hline \end{array}$

36

17

$\begin{array}{r} 49 \\ + 47 \\ \hline \end{array}$

27

Dino ahead. Wait 1 turn.

$\begin{array}{r} 20 \\ + 16 \\ \hline \end{array}$

Short Cut go ahead

96

$\begin{array}{r} 51 \\ + 25 \\ \hline \end{array}$

Chased by T. Rex. Move ahead 3.

$\begin{array}{r} 67 \\ + 42 \\ \hline \end{array}$

63

315

76

$\begin{array}{r} 34 \\ + 29 \\ \hline \end{array}$

109

$\begin{array}{r} 212 \\ +103 \\ \hline \end{array}$

Need a rest. Wait 1 turn.

$\begin{array}{r} 89 \\ + 79 \\ \hline \end{array}$

$\begin{array}{r} 215 \\ +212 \\ \hline \end{array}$

Spot a dino egg. Move ahead 3.

392

Short Cut go ahead

168

$\begin{array}{r} 256 \\ + 136 \\ \hline \end{array}$

427

48

$\begin{array}{r} 249 \\ +255 \\ \hline \end{array}$

Danger go back

Herd of dinos. Move ahead 4.

$\begin{array}{r} 29 \\ + 19 \\ \hline \end{array}$

349

Short Cut go ahead

504

See dino tracks. Wait 1 turn.

$\begin{array}{r} 224 \\ +125 \\ \hline \end{array}$

$\begin{array}{r} 39 \\ + 33 \\ \hline \end{array}$

Coast is clear. Move ahead 3.

896

721

692

72

$\begin{array}{r} 450 \\ +242 \\ \hline \end{array}$

$\begin{array}{r} 498 \\ +223 \\ \hline \end{array}$

$\begin{array}{r} 449 \\ +447 \\ \hline \end{array}$

389

Dinos fight. Wait 1 turn.

$\begin{array}{r} 297 \\ + 92 \\ \hline \end{array}$

Danger go back

Danger go back

end

23

Palindrome Number Search

"What's a palindrome?" you ask. A palindrome is something that reads the same backward and forward.

Words can be palindromes: **did noon radar eye**

Names can be palindromes: **Otto Hannah**

Sentences can be palindromes: **Madam, I'm Adam.**

Even numbers can be palindromes: **1221, 23432, 35,677,653**

Go on a palindrome number search. All you need is a number and a little addition.

Take the number 143 for example. It's not a palindrome but you can use it to find one. Just reverse 143 to make the number 341. Then add the two numbers together.

$$\begin{array}{r} 143 \\ +341 \\ \hline 484 \end{array}$$

Sometimes it takes a little longer. Just keep reversing the sum and adding until you find the palindrome.

$$\begin{array}{r} 48 \\ +84 \\ \hline 132 \\ +231 \\ \hline 363 \end{array}$$

• Use these numbers to find palindromes:

45   92   37   07   243   328   254

• Now for a toughie. Use the number 97. It will take six additions. You may want to use a calculator.

24

More Riddles and Puzzles

Use the decoder to solve the riddles. First find the sums. Then locate each sum in the decoder to find the letter it stands for. The first sum has been decoded for you.

DECODER

21	22	23	24	25	26	27	28	29
K	R	A	I	U	F	M	W	G

30	31	32	33	34	35	36	37	38
N	D	O	T	E	L	J	B	S

What did the dog say to the flea?

17	21	16	20		25	14	19		15	17
+14	+11	+14	+13		+12	+11	+10		+12	+17
31
D

What kind of snake is good at math?

19	17		13	16	23	22	15
+4	+13		+10	+15	+ 8	+12	+ 7

Why do birds fly south?

17	20		16	25		24	25	19		15	14	18
+ 7	+13		+ 8	+13		+ 9	+ 7	+13		+11	+ 9	+ 4

	19	21		16	13	21	13
	+14	+11		+12	+10	+14	+8

Addition Riddles & Magic Squares

When you finish this riddle, see if you can make up some of your own. Then amaze your friends with your riddlery.

I'm a four-digit number.

My ones' digit is the sum of my tens' and thousands' digit.

My tens' digit is the sum of my hundreds' and thousands' digit.

My hundreds' digit is 1 more than my thousands' digit.

My thousands' digit is 1.

What is my number?

Try solving these magic squares using decimals.

2.6		
	2.5	
2.8		2.4

		7.7
		8.2
7.9		7.5

Answers

P. 4, Birds on the Wing

You can find out how many geese are in each flock by counting. The flock on the left has 13 geese. The flock on the right has 11. To combine the two flocks to make one flock, you need to add the flocks together. You add to find how many in all.

You can add the flocks by counting. You know that there are 11 geese in one flock. You can think "13" and point to each of the 11 geese in the second flock and count "14, 15, 16," and so on. By counting the geese, you learn that 13 and 11 are 24.

You can also add the two flocks by writing an addition problem. Here are two ways to show addition:

$$\begin{array}{r} 13 \text{ geese} \\ +11 \text{ geese} \\ \hline 24 \text{ geese} \end{array}$$

13 geese + 11 geese = 24 geese.

If you add four more geese to your flock, you can find out how many there are by counting. Or, you can write another addition problem:

$$\begin{array}{r} 24 \text{ geese} \\ +4 \text{ geese} \\ \hline 28 \text{ geese} \end{array}$$

24 geese + 4 geese = 28 geese.

You can find out how many geese are on the pond by drawing a picture and counting. But it might be easier to use an addition problem:

$$\begin{array}{r} 28 \text{ geese in the flock} \\ +20 \text{ geese on the pond} \\ \hline 48 \text{ geese in all} \end{array}$$

P. 5, Be a Hat Checker

Everyone's tally chart will probably be different. When your chart is finished, you can count each of your tally marks to find how many. Or, you can count by fives since you know that each *HIT* is a five. Count on the last few tally marks if there are less than five. For example, if you had these tally marks:

HIT HIT HII III

You could count 5, 10, 15, 16, 17, 18.

You could count up all of your tally marks to find out how many heads you saw all together. Or, you could use an addition problem. Add the number of tally marks in the "Hats" column and the number of tally marks in the "No Hats" column.

Pps. 6–7, Number Shapes

Each side of the triangle has a sum of 12. Here are three different ways to arrange the numbers so they still equal 12:

Here are three ways to arrange the numbers 1, 2, 3, 4, 5, and 6 so each side equals 9:

Answers

This is one way to arrange the even numbers 2, 4, 6, 8, 10, and 12 so they add up to 18 (left) and 24 (right).

This is one way to arrange the odd numbers 1, 3, 5, 7, 9, and 11 so they add up to 15 (left) and 21 (right).

There are four addends on each side of the larger triangle. Each side adds up to 23. Here's how you can arrange the numbers so each side equals 20 (left) and 17 (right).

Each row in the wheel equals 10. Here's one way you can arrange the numbers so each row adds up to 12 (left) and 14 (right).

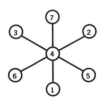

P. 8, Hit the Target
No answers.

P. 9, The Shortest Way 'Round

The perimeter of the rectangle is 28 cm. When you cut the rectangle in half along the dotted line, or diagonal, you wind up with two triangles. Each triangle has a perimeter of 24 cm.

Here are the perimeters of the shapes in the book:

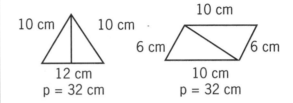

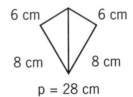

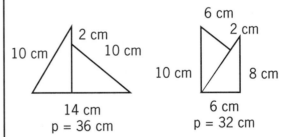

This is the shape with the shortest perimeter. It has the same perimeter as the rectangle you started out with.

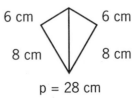

The longest perimeter is 36 cm. There are several shapes with that perimeter.

Pps. 10-11, Pyramid Magic

No answers.

Pps. 12–13, Magic in a Square

If you add the numbers in any row, column, or diagonal the sum is 15.

Your two magic squares should look like this. Their sums are 15 and 27.

8	1	6
3	5	7
4	9	2

12	5	10
7	9	11
8	13	6

You get these magic squares by adding 6 to each of the numbers in the squares above. The sums of these magic squares are 33 and 45.

14	7	12
9	11	13
10	15	8

18	11	16
13	15	17
14	19	12

The "Super Magic Square" looks like this. Its sum is 34.

7	12	1	14
2	13	8	11
16	3	10	5
9	6	15	4

The numbers in a magic square form a pattern. They are consecutive. That means the numbers run from the lowest to the highest without skipping a number. For example, the numbers in the magic square above are:

1, 2, 3, 4, 5, 6, 7, 8, 9, 10, 11, 12, 13, 14, 15, 16

Pps. 14-15, Four by Four

No answers.

Pps. 16–17, Coin Teasers

Here are 12 combinations for making 30¢:

quarters	0	0	0	0	1	0	0	0	0	0	0	0
dimes	2	1	3	0	0	0	0	0	1	0	1	0
nickels	2	4	0	6	1	0	1	2	0	3	1	4
pennies	0	0	0	0	0	30	25	20	20	15	15	10

There are many different combinations of coins for each price tag. Here are two examples for each one.

49¢	1	2	0	4
	1	1	2	2
37¢	1	0	2	2
	0	2	3	2
29¢	1	0	0	4
	0	2	1	4
79¢	3	0	0	4
	2	2	1	4
22¢	0	2	0	2
	0	1	2	2
85¢	3	1	0	0
	2	3	1	0
61¢	2	1	0	1
	1	3	1	1

Pps. 18-19, Sum Game!

The answers depend on the game.

Answers

Pps. 20–21, Zoo Maze

from the entrance to the monkeys= 94m

from the monkeys to the elk= 94m

from the elk to the gorillas= 86m

from the gorillas to the seals= 108m

from the seals to the camels= 132m

from the camels to the cafe= 63m

There are several ways to plan your tour. Here is one. It is 1,124m long.

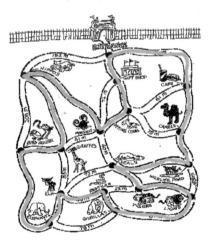

Here's another. It's 1,233m long.

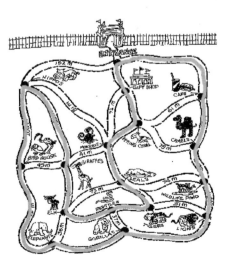

Pps. 22-23, Dinosaur Trek

No answers.

P. 24, Palindrome Number Search

$$
\begin{array}{r} 45 \\ +54 \\ \hline 99 \end{array}
\qquad
\begin{array}{r} 92 \\ +\ 29 \\ \hline 121 \end{array}
\qquad
\begin{array}{r} 37 \\ +\ 73 \\ \hline 110 \\ +011 \\ \hline 121 \end{array}
\qquad
\begin{array}{r} 107 \\ +701 \\ \hline 808 \end{array}
$$

$$
\begin{array}{r} 243 \\ +342 \\ \hline 585 \end{array}
\qquad
\begin{array}{r} 328 \\ +823 \\ \hline 1251 \\ 1521 \\ \hline 2772 \end{array}
\qquad
\begin{array}{r} 254 \\ +452 \\ \hline 706 \\ +607 \\ \hline 1313 \\ +3131 \\ \hline 4444 \end{array}
\qquad
\begin{array}{r} 97 \\ +79 \\ \hline 176 \\ +671 \\ \hline 847 \\ +748 \\ \hline 1595 \\ +5951 \\ \hline 7546 \\ +6457 \\ \hline 14003 \\ +30041 \\ \hline 44044 \end{array}
$$

P. 25, More Riddles and Puzzles

If you use the decoder to solve the riddles, you already know the answer.

P. 26, Addition Riddles & Magic Squares

The riddle answer is 1,234.

Here are the magic square answers:

2.6	2.7	2.2
2.1	2.5	2.9
2.8	2.3	2.4

8.1	7.6	7.7
7.4	7.8	8.2
7.9	8.0	7.5

Glossary

addend Any number that can be added to another.

example: $5 + 4 = 9$

addends

addition The process of adding two or more numbers together.

diagonal A line connecting opposite corners of a shape.

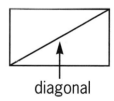

diagonal

digits The symbols used to write numerals: 0, 1, 2, 3, 4, 5, 6, 7, 8, and 9.

palindrome A word, name, or number that reads the same both forward and backward.

perimeter The distance around a shape.

predict To try and guess the outcome of something.

rectangle A shape with four sides and four right angles.

example:

sum The answer to an addition problem.

tally mark A mark used to keep track of items that are being counted. Tally marks are arranged in groups of five.

example: ⫽⫽⫽ ⫽⫽⫽ ⫽⫽

triangle A shape with three sides.

example:

31

Index

ML